my life cycle

Lion

Cherry Lake Press

Published in the United States of America by Cherry Lake Publishing Group
Ann Arbor, Michigan
www.cherrylakepublishing.com

Reading Adviser: Marla Conn, MS, Ed., Literacy specialist, Read-Ability, Inc.
Book Designer: Jennifer Wahi
Illustrator: Jeff Bane

Photo Credits: © AfriPics.com/Alamy Stock Photo, 5; © Viacheslav Lopatin/Shutterstock.com, 7; © Dagmara Ksandrova/Shutterstock.com, 9; © Maggy Meyer/Shutterstock.com, 11, 19; © Nick Greaves/Shutterstock.com, 13; © Krill Dorofeev/Shutterstock.com, 15; © Tanya Keisha/Shutterstock.com, 17; © GUDKOV ANDREY/Shutterstock.com, 21; © Dave Pusey/Shutterstock.com, 23; Cover, 2-3, 6, 18, 22, 24, Jeff Bane

Cherry Lake Press is an imprint of Cherry Lake Publishing Group.

Library of Congress Cataloging-in-Publication Data

Names: Gray, Susan Heinrichs, author. | Bane, Jeff, 1957- illustrator.
Title: Lion / Susan H. Gray ; illustrated by Jeff Bane.
Description: Ann Arbor, Michigan : Cherry Lake Publishing, 2021. | Series: My life cycle | Includes index. | Audience: Grades 2-3
Identifiers: LCCN 2020030675 (print) | LCCN 2020030676 (ebook) | ISBN 9781534180062 (hardcover) | ISBN 9781534181779 (paperback) | ISBN 9781534181076 (pdf) | ISBN 9781534182783 (ebook)
Subjects: LCSH: Lion--Life cycles--Juvenile literature.
Classification: LCC QL737.C23 G7175 2021 (print) | LCC QL737.C23 (ebook) | DDC 599.757--dc23
LC record available at https://lccn.loc.gov/2020030675
LC ebook record available at https://lccn.loc.gov/2020030676

Printed in the United States of America
Corporate Graphics

table of contents

About the author: Susan H. Gray has a master's degree in zoology. She loves writing science books, especially about animals. Susan lives in Arkansas with her husband, Michael, and many pets. But no lions.

About the illustrator: Jeff Bane and his two business partners own a studio along the American River in Folsom, California, home of the 1849 Gold Rush. When Jeff's not sketching or illustrating for clients, he's either swimming or kayaking in the river to relax.

I'm a baby lion only a few hours old. I have a thick coat with dark spots.

My eyes are closed, and I can't see. But I feel my wiggly brothers nearby. We are **newborns**, drinking our mother's milk.

My family lives in a group called a **pride**. Most other lions here are my female **relatives**.

Our aunts and grandmothers often look out for us.

Soon, my eyes open and my teeth appear. I'm a young cub now. I practice my roar.

I like to **explore** and **wrestle** with my brothers.

As I get older, my spots fade. My brothers start growing their **manes**. We are becoming **juveniles**.

We learn to hunt by watching our mother.

Our pride includes a big male with a dark mane. He becomes my **mate**.

About 4 months later, I have my own little cubs. We **snuggle** close.